Sin & Scientism

SIN & SCIENTISM

Jacob Needleman

ROBERT BRIGGS ASSOCIATES
SAN FRANCISCO

Published by
Robert Briggs Associates
Box 9
Mill Valley, California 94942

Editors: Nancy Kleban, Kathleen Goss
Design by Mark Ong
First Broadside Edition 1986

ISBN 0-9609850-7-7

Introduction

Since 1972 Robert Briggs Associates has been concerned with the publication of work related to the often confusing field of human transformation, an area where the separation between the sublime and ridiculous is not always apparent. The wave of interest in this subject, thought by enthusiasts to be a sign of a New Age and by many critics to be mere narcissism, has been both promising and disturbing. Whatever the impression, the current transformation in America has been the result of change no concerned individual can ignore. Its ramifications have been widespread and are dramatically exemplified by the effects of quantum theory on science, of holism on medicine and psychology and of Eastern beliefs on Western religious ideas. Willy-nilly, all these are influencing the way we perceive reality, and hence our images of ourselves. We are changing whether we like it or not.

Because of the eclectic nature of transformation and the sometimes eccentric character of its champions, publishing problems have always been pronounced. The intrinsic interest and worth of such works as Doug Boyd's *Rolling Thunder,* Arthur M. Young's *The Reflexive Universe,* Kenneth R. Pelletier's *Mind as Healer, Mind as Slayer,* or *Beyond Biofeedback* by Elmer and Alyce Green are readily apparent, yet speculative consideration of the potential for human transformation—eventually seen by many as an increased awareness of the nature of consciousness—often produces commentary and criticism too short for a traditional book but too long or unusual for the commercial magazine, specialized journal or anthologized mix.

It became apparent some alternative was needed. Although the first Broadside Edition was not published until 1985, the concept, considered in 1981 as a vehicle for written work that did not fit traditional moulds, inspired over a hundred themes, as well as a variety of complaints and some possible celebrations.

From the beginning the relationship between sin and scientism aroused interest. There was a definite correlation between contemporary ethics—our secular interpretation of right and wrong—and scientific excesses in the New Age. But that correlation, important as it seemed, was difficult to expose or explain. Most professionals were either uninterested in or unqualified to discuss more than one discipline. Too few psychologists were aware of the implications of Eastern spirituality upon Western lifestyle. Fewer physicians seemed concerned with the growing ethical problems brought on by technology. It was difficult to find scientists at all qualified to discuss philosophy, or philosophers who had the necessary understanding of quantum theory and the revolutionary implications of the Uncertainty Principle. (Of course, in the past thirty years, there have been exceptional pioneers and signs of remarkable disciplinary diversification. The work of Abraham Maslow in psychology, Gregory Bateson in anthropology and countless others—Colin Wilson, Joseph Campbell, Stanislav Grof, Theodore Roszak—has had historic implications, and recently the efforts of William F. Thompson, Fritjof Capra and Rupert Sheldrake have added to this enlightenment.)

The correlation between sin and scientism was broached early in 1984, when Jacob Needleman spoke at the Institute for the Study of Consciousness in Berkeley, California, and touched on "the encounter of modern science and ancient truth." He had previously examined this topic at some length in *A Sense of the Cosmos,* a work dealing with "modern man between two dreams" and including chapters on medicine and physics, as well as on psychotherapy and the sacred. In the spring of 1985, with the publication of *The Way of the Physician,* a book the New York Times said made us doubt whether "technology alone has the power to save our souls," Jacob Needleman again proved to be one observer capable of dealing with sin and scientism.

I

Robert Briggs: Perhaps the first question should be, What is sin? And then, What is scientism? Perhaps the question should be, Is scientism a sin? Or, Is there a point when scientism becomes sinful? The consideration of sin and scientism seems to rise out of a growing concern with the quality of life. We are constantly reminded of the wonderful new age we inhabit with its future of space stations, Star Wars (that will supposedly make the cosmos safe for democracy), computer technology, genetic engineering, mega-trends, mega-celebrities, and all kinds of opportunities. Yet despite the wonder, it seems impossible to escape the gravity of modern problems. Despite a New Age promise, there is haunting discontent in most of us that's never satisfied by the latest scientific breakthroughs that seem to come with suspicious regularity.

This discontent feeds on distrust and doubt about such things as the horrors of world hunger, world debt, pollution, terrorism and the nuclear arms race. It seems to come out of old tyrannies—Marxism, behaviorism, religious extremism—out of old morality that seems so dated and out of new moralities that often seem all too conveniently adjusted. So it's not difficult to stand back and find a common root in sin and scientism or a correlation between the two. Indeed, it is only too clear that science, and especially what is called *scientism*—the elevation of science to the status of a guiding principle or philosophy of life—has something very fundamental to do with our predicament. Is there a point at which scientism actually becomes *sinful*? Can we talk about ethics and science?

Jacob Needleman: Well, yes. Ethics is, strange to say, a very neglected subject in the modern era, even though everybody speaks about ethics nowadays—business ethics, professional ethics, medical ethics. Ethics is really the study of human relationships against a metaphysical background of purpose and meaning in the universe. Whereas psychology studies human relationships more or less against a background of individual pleasure and pain and emotional satisfactions. So ethics, to really be ethics, has to be set in a cosmic or even metaphysical framework. And science obviously is our main instrument in our modern era for telling us what the universe, what nature, is really like. So our ethics, or lack of ethics, is really a direct result of how our science is conducted. If our science has gone sour, then our ethics inevitably go sour.

RB: In *The Dictionary of Philosophy,* Antony Flew says that sin is "strictly a theological term, meaning offense against God. What is sin, therefore, may or may not violate worldly moral standards."

JN: I don't agree with that definition.

RB: What would yours be?

JN: Well, that sin is a wrong against God is all right. It's fine to say that, but then God is not just some removed figure up in the sky but is the ultimate, principal reality in the universe. So sin is not strictly a theological term at all. I don't think you can separate the two. If you make a distinction between worldly and theological, then you're fine. That means that the standards people generally adopt to get along are different from the ones that they adopt in relationship to the ultimate principle of reality. Which means, basically, that our standards for right and wrong are cut off from reality.

RB: When this is discussed with certain critics, they are quick to point out that the Buddhist rejects or ignores the concept of sin, inferring that sin is a Christian responsibility.

JN: Well, if you define sin in terms of narrow Christian dogma, perhaps. But Buddhism does not reject sin in any meaningful definition of sin, which is: not acting according to what is right and good in ultimate terms. Buddhism has a very definite notion of sin. The Boddhisattva ideal is that of an individual person who is doing great good, is deeply moral, is helping others out of their suffering. And the opposite of morality is sin, in the more general, meaningful definition of the term. So yes, if you want to limit it to Christian theology, then we have to find another word. But I think the word is too rich and strong to limit it to just doctrinal things.

RB: Today elements that control our lives, and the fate of the earth, seem to be above or beyond a code of ethics. For instance, what about the ethics of the American banker? Many contend that the problem of world debt is directly related to the fact that the American banker has enmeshed the Third World in debt the Third World cannot hope to repay. On the other hand, a part of the economic well-being of America is based upon these debts. Is there an ethical problem here? Is it right that the American banker has extended this kind of credit?

JN: You can't say yes or no. It's an impossible question to answer. Ethics does enter into it, of course. As long as we're talking about relationships between people, ethics enters into it. I don't think you can say a banking system as such has ethics. I don't know if that would apply.

RB: Of course, no system has ethics as such, but what about the individual banker? Or the corporate executive who moves steel manufacturing from Pennsylvania to Korea because it's good for the corporation, when such a decision leaves an entire area of America impoverished? Is there an ethical problem for the corporate executive?

JN: Well, he might justify his decision by facts and information and ideas which you or I might not agree with, or even know about. We must remember that no banker or executive or anyone does bad consciously or intentionally. This brings to mind Socrates' famous statement that nobody is intentionally evil. Everybody does what they consider to be good. It's their understanding of it that's different. So the banker or the corporate executive could very well tell you: well, if it brings a profit to the company, if it helps make better products available, ultimately it's going to be good for America and good for Americans. The point is whether or not ethics is thought about in this context—not what's right and wrong. That's very hard to judge from standing outside. But how does ethics enter into this behavior at all? Does the banker even stop to think whether it's good or bad? And I think he does. Nobody wants to go through his life thinking that he's completely without ethics. I believe the banker does think about that sort of thing.

RB: This brings to mind an impression I got from your essays in *Consciousness & Tradition*. There you imply that the disintegration of ethical or moral standards on any level is due to the loss of, or disintegration of, a kind of spiritual skill.

JN: Yes. My point of view would be that we're all enmeshed in forces we don't understand. The banker, the corporate executive, no matter whom you speak of—the rich, the poor, the privileged or the disenfranchised—everybody is at the mercy of forces they don't understand. Nobody can really help what they're doing. Now that's putting it very bluntly. The question is: how do we understand those forces? How do we study them? The spiritual skill you're referring to begins with understanding how these forces operate in us.

You know, in the New Testament Saint Paul defines sin by saying, "The good that I would, I do not: but the evil which I would not, that I do." And he goes on to say that it's as though there were forces operating in me that are not me, but which take me over. And they were symbolized by devils in earlier eras. But it's the same point. We have in us a voice that wants us to do something that's good, but there are forces operating in us that make that difficult.

RB: But wouldn't it be the consideration or revival of spiritual skill that would bring back an ethical standard?

JN: I think so.

RB: And in this respect should the banker, executive, or physician be responsible? Or do you think we're all swept away? And do you see this as a failure of modern Christianity? In other words, has Christianity been overwhelmed by these economic, material forces?

JN: I think most followers of religion are just like the rest of us.

RB: Yes, then let's go a bit further. Do you think Christianity itself—always a source of our ethics—has been overwhelmed by Eastern arrivals? By the guru, the swami, the baba? Has contemporary Christianity been overwhelmed by what some people consider a terribly antiquated Pope trying to reverse the modernization of Catholicism and hindered by Moral Majorities that are far more political than spiritual? Is this a failure of Christianity? Christianity, of course, not being the only American religion, but certainly the one which most bankers and executives come from.

JN: The pertinent question is the precise extent and nature of the failure of religion in our time.

RB: Of course the defender of the faith would say: well, look at the ministers who marched in the South during the war for integration and look at the ministers who are out in the forefront of the struggle against the ecological crisis. But you can always set that against the number of impoverished or deserted churches or the number of clergy and nuns who have left the Catholic church in America.

JN: I think the principal problem with the Christian church today, as in other Western religions, is the loss of contact with the spiritual, contemplative dimension. I don't think it's possible to be deeply ethical unless you have access to the deeper layers of the self, and such access is only possible through the practice of an authentic spiritual discipline, and I think the church, whatever it does, with whatever intentions, as long as it doesn't offer access to a precise approach to the deeper layers of the self, can never help people to become deeply ethical.

RB: It seems that if one wants to seek this in America today, one really has to turn to Eastern religions. What with the popularity of meditation.

JN: In any case, that's a sign of the attempt to return to a real, nourishing source of spirituality. The Eastern religions, where they are authentic—and I'm not talking now about charlatans or fake gurus—where the Eastern religions are authentic, they point to a mode of access to deeper layers of the self which the Christian religion has lost contact with, and which many Christians feel needs to be rediscovered. But you must remember Christianity itself has a long tradition of this kind of meditative discipline, but it has not played an important part in church Christianity for many centuries.

RB: Do you think the Eastern arrivals—the guru, the swami, or their ideas and ideals—will be swept away by the power of Western materialism? I remember—I don't know if you want to re-tell this—the time you escorted what you considered a very serious Eastern wise man to a speaking engagement. He gave the speech, he wowed his audience, and then you were to drive him to the airport. What happened then?

JN: Yes, that was interesting. He wanted to cash the honorarium check, so we stopped at a bank and he presented it to the bank teller, who turned out to be a very timid Japanese American woman who said, "I'm very sorry, but we can't cash this check if you're not a citizen." And he flared up and shouted at her so angrily, because he was late for his plane, that she practically flew against the wall. When she went scurrying after the bank manager, I said to him, "We can talk wonderfully about spiritual values, but it's when we're trying to cash a check in a hurry that our values really are tested, aren't they?" And he admitted, "Oh yes! yes! yes! yes!" But it didn't really penetrate. The man was furious.

RB: Well, it's very easy to be cynical. In the past ten years I've watched one impressive swami who arrived in America weighing 145 pounds. His psychophysiological powers were incredible, but ten years later, when he controlled a million-dollar ashram, he weighed more than 200 pounds. It's said he doesn't even meditate anymore.

JN: We can't be too optimistic about our swamis. Many of them get swept away as quickly as any of us, priest, rabbi or corporate executive. Being called a swami or yogi doesn't ensure that you're not going to get swept away. I'm just saying that in general the Eastern part of the world has lost less of the inner disciplines. If we could get enough authentic representatives of these disciplines, it might help, but many swamis who have come here are probably as vulnerable as anyone.

RB: In a sense, aren't we comparing more than religions? Aren't we touching upon different kinds of disciplines? Different methods coming up against one another? And doesn't the domination of scientific method in the West, or scientific materialism, hinder the development of more "inner" spiritual skills? Can they ever coexist in the modern world?

JN: I think the scientific discipline is a noble ideal of acquiring external knowledge, without being fooled, without subjectivity—you test what you believe, you put it to experiment, you verify it, and you see if it works; you study its practical consequences. But in general—except for rare flashes of deep insight—scientific work makes use of only one small part of the mind. Or you could say it is a channel of only one kind of force in human beings. But let's speak about it the first way: it makes use of a part of the mind which functions in a certain way, very often like a good computer. Whereas the spiritual discipline we're talking about is a mode of access to another, deeper part of the mind, and those two can exist at the same time in our lives, but one has to see that very different parts of the mind are involved here. Now, ultimately, they have to be brought into harmony with one another, but before they can come into harmony, it is absolutely imperative that they be seen as different, and not too quickly pushed together or "synthesized."

RB: Not unified with economics or aerobics.

JN: Right. Not unified with aerobics or in any cheap way with one another. The harmonization of these two levels of mind is accomplished only at enormous cost. It's very difficult to unify these two. One of the main difficulties is that in the West we are impatient to bring things together before we have really studied their differences.

RB: We're desperate.

JN: We're very desperate, but perhaps not desperate enough. If we were really desperate, we might take our time to realize and see that these are difficult things. When you're truly desperate, you can be very patient.

RB: So, in considering sin and scientism in America—where our lives seem dominated by scientific authority—is it possible to speculate that something's got to be done about raising the question of right and wrong with regard to scientific materialism?

JN: Absolutely.

RB: Obviously, shaking fingers about sin is not going to move bankers or executives. Still, it's a question of ethics. You seem to infer that the individual has to come back to the self.

JN: There's no ethics without the self. A computer can't be ethical. A mechanical human being can't be ethical. There's no scientist who can be ethical if he's not in touch with this inner world. There's no executive, no writer, no Aquarian person who can be ethical unless they are in touch with the part of the human being which is free, which is alive, which is able to make choices. So the whole question of ethics may be a complete—what shall we say?—distraction, until you are in touch with the parts of the self which can be ethical.

RB: This calls to mind the practical problems of unification of the psychological and the physical self—psychophysiological well-being. One of the difficulties is that the individual seldom realizes that fitness and revision of diet alone will not produce the removal necessary to get in touch with the inner self. There has to be a third element, some kind of meditative relaxation.

JN: Oh, and many other things, but certainly that, yes. There have to be many, many things, because the human being is a very complex organization, and many things are wrong, and you can't just right one or two of these wrongs and hope to correct the whole person. Many popular groups and movements do very well touch one or two problems, and whether it's aerobics or psychotherapy or Transcendental Meditation or any one of a number of things that are being offered—particularly in California—they do touch one or two things that are wrong with us. But the point is to touch them all. Just correcting one or two things is not going to do the job. We're too complex, and the damage that's been done to us is too embracing. It goes over the whole of ourselves. That's my view. I don't think it's pessimistic, but addressing the whole thing is much harder and requires a much greater knowledge than most of us realize.

RB: Let's go back a little. Antony Flew calls scientism "1. The belief that the human sciences require no method other than those of the natural. 2. In a more general sense, practices that pretend to be, but are not, science."

JN: I don't know if I followed that. Scientism is a belief that you can understand everything with the parts of the mind that are used in ordinary, everyday natural science.

RB: Well, then, let's try the *American Heritage Dictionary.* There scientism is "the theory that investigational methods used in the natural sciences should be applied in all fields of inquiry."

JN: All fields of inquiry, and all fields of endeavor, then you have scientism, yes. And that is absurd.

RB: That *is* absurd. If scientism is applied to philosophy, you have no philosopher.

JN: Right. Philosophers have sort of put themselves out of business, because they have on the whole, in America and England, assumed that science is the only real, reliable access to knowledge or the use of knowledge. The only real kind of knowledge that is reliable. Yet there are enormous parts of the world and human life, perhaps the most important parts, which are inaccessible to the part of the mind that is generally employed in doing science.

RB: Let's come back to the philosopher and go on to the scientist and say something about the ethics of the scientist. The point has been made that we've reached the end of a scientific era that began over 300 years ago with Bacon and Newton and culminated with Einstein and Heisenberg. The lineup is deservedly famous. In this era science was unleashed and its success historic. The technology it produced—or the technology that evolved because of it—produced the "better living" we enjoy today. But as this century comes to a close, the scientific dazzle seems to have short circuited. Someone said that the technological showboat has "dun run aground!" Despite all assurances, science has poisoned our food, our water, and endangered the very atmosphere in which the scientist, as well as the individual, lives. One critic wonders whether or not present technology creates more problems than it solves. Nuclear waste is a perilous example. Nevertheless, we are still enthralled, if not captivated, by the overbearing insistency of scientific urgency. An urgency that's become so persuasive that it's led to a fateful suspension of whole, raw life. The life of the mind, the life of the spirit seems postponed. That life in which a sunrise or sunset is as rich and as important as another, larger, faster rocket to the moon seems suspended. Perhaps the question is this: If modern science has reached a limbo, isn't it the ethical responsibility of the individual scientist to question the extent of scientific authority? Traditionally, the scientist has claimed he's only responsible to science. But what about the scientist destroying himself? Shouldn't we ask the scientist to step back and reconsider the ethics of this?

JN: We'd have to help him step back. But who are we speaking of here? When you say the scientist, who are we speaking of? Most of the population, in one way or the other, in the Western world is involved in science or technology. Most of the best jobs are in science or technology. So we're speaking of a small minority of all these people—the best and the most influential scientists. And some of them *are* trying to step back. But I'm not sure they know exactly how to step back. One of the hardest things in the world is to step back from one's presuppositions. And that used to be the function—spiritual function, if you like—of the philosopher, to help human beings step back from their presuppositions and become aware that they are only presuppositions. And the philosopher, apart from some of the even more noble aspects of philosophy, could have a role to help the scientist become free of his presuppositions. I don't think you can expect it of the scientists themselves. They're like you and me. They know things have been going crazy. They know that every time you make a technological improvement, you create two new problems for every solution.

RB: They might be more frightened than we are.

JN: They probably are. And they know they don't know enough. They know that nobody knows enough. They're overwhelmed with information, with data, with theories, with discoveries. There's absolutely no one who can have an overview of the whole thing at this point. So it's not only run aground, it's just gotten out of control. It's just shooting all over the place, and nobody is in control.

RB: It's our inheritance of the past three hundred years. In *Forgotten Truth,* Huston Smith claims that the modern West is the first society to view the world as a closed system, but he says that view, in principle, is impossible, for "taken in its entirety, the world is not as science says it is; it is as science, philosophy, religion, the arts, and everyday speech say it is." So if we're having the trouble with scientism that we fear we are, is it too idealistic to wonder whether or not the scientist should be seeking the philosopher?

JN: No, I don't think you can ask that of the scientist. I think the philosopher has to go out looking for the scientist. I think that the moral obligation of the philosopher is to go after the scientist. Why should a scientist go out to look for a philosopher? When they find the philosopher, all he turns out to be is somebody who worships the scientist.

RB: What happened to the philosopher? When I was in college in the middle 1950s, there was a philosophy department. Now philosophy departments seem even more remote. Is this because of questions philosophers could have asked, or should have asked, and didn't?

JN: On the whole they've bought into the scientism thing too completely. Whereas the scientist every day is discovering the limits of science, the philosopher is not out there on the front lines discovering for himself—painfully—that science is really screwed up in many ways. So the philosopher bought into the scientism model. That is to say, he began to assume that the kind of thinking and the part of the mind that science uses is the most reliable, best source of knowledge about everything in the world. So philosophy, putting it simply, became, since the early 1940s, basically a handmaiden of science.

RB: You're talking about specific schools of philosophy?

JN: General, academic, mainstream philosophy in the United States and England.

RB: In other words, there are not enough people complaining about the loss of spiritual skill?

JN: No. Nobody knew how to complain about it. Those who did complain about it didn't have the language to make an impact on the scientism thrust. They were using old language, the language of Christian spirituality, which nobody wanted to listen to. Like the word sin, which we need to redefine. That old language was dead. Nobody was going to listen to someone like Jacques Maritain saying we've got to have spirit. No philosopher, no scientism person is ever going to listen to that, even though, as I think it is meant, it is true. But you have to redefine it so that the scientist can listen. You have to help people to listen, not only tell them what they should hear. You have to help them unplug their ears so that they are able to hear. Plato was very right. He said one of the most important aspects of the true philosopher is what he called rhetoric—not in the modern sense, but in the sense of the art of communicating great ideas.

RB: Do you think in some ways the philosopher—in talking about these forces we've all been swept along with—do you think the philosopher was swept into this position?

JN: Absolutely, completely.

RB: The philosopher lost his societal importance, and the loss of academic importance followed. Universities cut the funding of philosophy, because philosophy had lost its importance.

JN: Well, it was some of that, but as I said nobody is intentionally evil, nobody is intentionally materialistic, in the sense of having no feelings of right and wrong.

RB: No, but people get very political when it comes to material or academic survival.

JN: Yes, of course, and it's a big force. But let's put it this way—without judging whether they have economic motives or other motives—that they got swept up into the triumphant surge of scientism, starting in the 1940s particularly. They became philosophers of science, basically.

RB: Yes, but in the 1980s, experiencing the fallout of technology, we suffer this terrible discontent. Shouldn't the philosopher be out on the forefront asking tough questions?

JN: Yes, but you can't expect that of an entrenched profession. Many real philosophers are coming out of other places now.

RB: Such as?

JN: Well, sometimes you find them in literature; sometimes, very often, in the religions. In the sciences, often physicians. All kinds of people are asking philosophical questions. We haven't yet got the real philosopher on our scene who is going to give us a world view that is noble, all-inclusive, coherent and compelling, and which the scientist, the artist and the person in religion can agree to.

RB: Here in San Francisco you've been called upon to speak to executives and physicians about the problem of ethics. Are other philosophers being called upon?

JN: Yes, sometimes, but I'm afraid that often it's in a much narrower sense. Doctors, for instance, will call upon an ethicist in order to help them think through a difficult medical situation rather than to help them toward a vision of something bigger, deeper, in themselves and in the universe.

RB: That's an interesting term: ethicist—it's almost public relations. In a sense another part of the scientific excuse.

JN: Right. Ethicist is a rather ugly term.

RB: Yet there's a whole profession coming out of it.

JN: I know. Like the professional lover.

RB: Some corporations now have their in-house ethicist.

JN: Yes, but that's not what we're talking about. In that sense ethics is simply the art of covering your ass, and that's not what we're talking about. What we're talking about is opening your heart rather than covering your ass.

RB: A matter of educating a part of the person that isn't being educated these days.

JN: Absolutely.

RB: Shouldn't the philosopher have some renewed role in this education?

JN: Well, if philosophers could recognize this part of the self, yes. But they don't do it yet because they're mostly scientifically oriented. So you can't expect that of them. They're not into that, on the whole. But there are other people who are. You just can't tag them. So you have to find people who are sensitive to this other part of the self which Plato knew about, which Spinoza knew about, which all the great philosophers of the ancient world knew about, and which somehow most of our philosophers don't know about, except in words.

RB: In other words, they're suffering from what I think you called the "automatization of humankind."

JN: Right.

RB: And the philosopher— when this automatization takes place—is drawn away from true philosophy.

JN: Absolutely.

RB: Right into scientism, which we fear is what science has fallen into.

JN: Absolutely.

RB: So let's go back to our poor concerned individual out there in the world. Intelligent, very willing to listen and read—very willing to face up to modern problems and haunting discontent. She or he instinctively knows something is wrong. But what's to be done? Don't they need a spokesperson? Somebody to cry out, a philosophical Ralph Nader, somebody to create that shock of recognition of what is so wrong. Have we reached a point where we can't dramatically change trends?

JN: Well, maybe. But it's part of the illusion of our time that we're going to make evolutionary changes on a mass cultural level. I think the Ralph Nader syndrome is that you've got to get a big mass reaction and that will transform the direction of our whole civilization. I don't think that's the way deep change is going to come. It's going to come, much more likely, through individuals finding each other and working hard together, quietly, behind the scenes if you like. That can radiate an influence on the whole society. If you got a philosophical Ralph Nader, it would only be another turn of the wheel of *samsara*—more grist for the mill, more of the same old thing.

RB: Are you talking about a neighborhood level?

JN: Neighborhood isn't the unit I'm thinking of. It can be across the world, but people who have in common that they are searching for something that even the best of ordinary life can't give them—these people usually recognize each other.

RB: Somebody made the point that in so many ways the new age tends to isolate individuals rather than integrate people. It isolates because it compartmentalizes.

JN: Yes, that's a danger. I agree that what one needs to do is to get more involved with the world.

RB: If you were talking to one of those concerned individuals and that person turned to you and asked, "What should I do? This afternoon, this evening or tomorrow. How should I begin?" What would you say?

JN: The first thing I'd say is, "Do you have more than a moment?" And I would like to spend some time talking to the person. There's no simple solution—no going out to march for the higher self, or something like that. That's not the point and you're not asking that, I don't think. What should a person do? Well, let's hear him! What is she or he saying? What's his problem? What's her question? What is *the* question? You say discontent. Everybody is discontented. They always have been.

RB: In other words, the variations might defy formula.

JN: There is no formula, I don't think. It's a question of relationship. Can I, or you, relate to another person in a search for truth—not in an attempt at a quick fix or some immediate satisfaction, or some illusion, or whatever? Can we relate to each other in the quest for truth, even if it's a small truth? That is ethics—relationship under the light of truth. Now, if this person wants to inquire deeply, I would like to meet him. I would like to meet and talk. I'm that person. You're that person. We all want to know. These things take time. You can't say to a person out there who's discontented, "Here's what you should do." You need to go into things together. It takes time. Not too much time, but it takes time. You cannot get to this—snap!—like that. That's part of the problem we're all facing. Even discontented, intelligent, serious people feel they've got to find something right away that's going to work. It took a long time for us to get this screwed up. It's not going to get fixed up with one or two things or one or two days or months.

RB: Do you consider meditation to be valuable?

JN: I think meditation's indispensable.

RB: We're at a point at which the individual has got to do something with the self before doing something for the self.

JN: A good way of putting it.

RB: Meditation or some kind of meditative relaxation. Unfortunately, at this point, when we mention meditation, we probably turn off 75 percent of the concerned Americans who associate meditation with something Eastern or voodooistic.

JN: That's because there are a lot of weird things being done under the name of meditation. And the word meditation by itself doesn't communicate enough. Some kind of work with the self—how did you put it?

RB: Doing something with the self before you do something for the self.

JN: And that requires knowing exactly what to try. Meditation is not a freelance enterprise. It's like somebody who wanted to do some medicine. You can't do do-it-yourself brain surgery—far less do-it-yourself work on higher parts of the self. You have to have knowledge. You have to know what to try and how to try it.

RB: You're saying this requires a teacher.

JN: It requires some guidance, I would say.

II

RB: We should talk more about spiritual skill. About what the concerned individual does. With people playing larger and more diversified roles in our society today, we can all share a common level, another plane of awareness, which is sort of the Boddhisattva ideal you mentioned earlier. One which can be communicated across disciplines. I'm wondering how that desire can be aroused. Once people become aware of the potential of awareness, it's easy enough for them to move society in that direction.

JN: Well, I'm not sure it's so easy even when the desire is aroused, but certainly without the desire nothing is possible. You have to ask yourself how this kind of desire is aroused in your own life, your own self. The only way you can answer that question about others is to know how you yourself came to want this sort of thing. And that itself is already the beginning of spiritual skill, which is to study what it is in life that brings you to this question.

RB: Do you think greater awareness of the cause of haunting discontent might bring this about?

JN: It could. But the main thing is that *I* become in question—me, I, right here—not them. When *I* am in question, when *I* am feeling that the whole of my life has been built on sand, when *I* feel that I haven't really understood why I'm here, or what life is—what my life is—then, at that very moment, I'm on the spiritual path. So we have to ask ourselves what brings us into question. We have to study that. And then begin to cooperate with moments when I am myself in question. This is the real beginning of the spiritual way for us and, I think, for everybody. But we can't pretend to apply things to others until we have studied ourselves.

RB: Since the root of awareness is knowing, perhaps it's time to talk about that. You've said that there are four kinds of knowing, four levels of knowledge.

JN: I would start from the bottom, which I would say is data. On the next level would be information, and the next level up we could call knowledge, and the highest level we could call understanding.

(1) Data
(2) Information
(3) Knowledge
(4) Understanding

All four can be called, loosely speaking, knowledge, and they have been, but more precisely they're very different levels of knowing.

RB: The same thing?

JN: It can be the same object that you're knowing, but they employ different kinds of material, different kinds of perception. Most importantly, a different part of the mind is engaged.

Data is simply the receiving of raw sense perceptions such as figures, statistics, reports—the bare, simple looking at something and reporting what's there, as far as you know.

Information is when this data is organized into patterns, such as the population of one country that gets sick from something more than the population of another country. You take data and organize it into patterns, so that you have some sense of facts. I think information is the world of facts. Facts are the organization of data, if you like.

The next level is knowledge, which is often theoretical, hypothetical, with concepts and ideas generalizing all your information under a simple and maybe brilliant synthetic unity of some sort. Very often this is something which a great scientist, or even any individual in his moments of insight, can have into the connectedness of things. And this is really a very important, high function of the mind. But the highest one, understanding, is something which is hard to define.

Understanding is a direct awareness of reality that's not simply mediated through thought, concept, but through another faculty, another ordering of the mind, which we very badly use the word intuition for. Direct apprehension of reality—really an understanding in the sense in which you can live and act and behave according to what you have seen to be true. You can have knowledge and not be able to live according to it. But understanding—maybe another word for what I'm calling understanding is wisdom, but that word might be too loaded—understanding is the highest kind of knowledge. A knowledge in which you see directly what is the real, and which you can incorporate in your living.

Those are four levels. I don't know if I've defined them enough, but those are four levels of knowing. Now when we say in the modern world we know more than the ancients did, that's very misleading. In a way it's stupid because we what we have is more of (1) and (2), more data and information, but, in fact, probably not much more of (3) and incomparably less of (4).

RB: Sometimes one gets the feeling that knowledge and understanding have been purposely eliminated. They upset a material view.

JN: Certainly understanding. It's very hard to find anybody who does that. There are still knowledgeable people around, but data and information are what is so common. Now, the ancients, masters if you like, the great guides of mankind over the centuries, didn't place all their trust in data and information. In great traditional civilizations data and information were not given exaggerated importance. Although they were always recognized as necessary, they were not given the honorific name of knowing and understanding. It's true that we know more than the ancients, but that only means we have more data and information. We have far less of the third and fourth, taken together. What I think is so interesting and part of the crisis of our time, is that there are questions and problems of living, very central questions, which can only be approached through knowing or understanding, particularly through understanding. And we are trying to solve these important questions with data and information. We never will. We will never solve the problems of ethics and living through data and information.

RB: Or seriously consider sin or scientism with data or information.

JN: No, you can't look at questions that require knowledge and understanding, when all you're using to solve them is data and information. That won't help in the slightest. So part of what you call haunting discontent comes from the frustration of people saying, "We know so many things, how come we can't live well?" Well, we don't really know or understand that much. We only have greater amounts of data and information. People ask, "If we can send a man to the moon, why can't we behave ourselves?" Well, behaving, living well, living rightly—

RB: —reducing crime—

JN: —reducing crime requires understanding of what crime is, not just data about where crime is being committed or information about what are the social-economic patterns of all this sort of thing. We can send a man to the moon because that requires a little bit of knowledge and a lot of information and data. But reducing crime requires a great deal of knowledge and an awful lot of understanding. So if we're not using the right tools for the problems we're so upset about, we'll never solve them. You can have endless amounts of information and endless amounts of data, and you can pile it up and on—there's an infinite amount of data about the very room we're sitting in—but by itself it won't give you the slightest bit of understanding.

RB: It would seem then, if we're not being too idealistic, that this would be an inspiration to the philosopher, the remote philosopher or the philosopher teaching in a college or a university. It would seem that this would provide the opening, the inspiration to confront scientism, and revive the advantages of spiritual skill.

JN: Yes, but only if philosophers understand spiritual skill. Without that understanding, they might agree to what you're proposing, but then they would attempt to confront scientism using the kind of thinking that comes from scientism itself. And that's where it all goes wrong. We philosophers may agree to something, verbally, conceputally; that doesn't mean we have understanding of it. You've got to find people who've got this sort of spark, this hunger, this yearning, who are themselves in question. And that can come from within spiritual traditions, it can come from the arts, it can come from philosophy, it can come from science. It's not limited to the categories or disciplines that we make up. The real philosophers are not necessarily the ones teaching in philosophy departments. There may be somebody out in Zaire working with a native tribe there who is a real philosopher. There may be an Emergency Room physician who is a philosopher. A philosopher is somebody who loves truth. That's the definition. Who loves wisdom—that is, who loves knowledge number (4).

RB: Was this less of a problem when you were at Harvard in the 1950s?

JN: Oh, no, it was a horrible problem then.

RB: As bad, or worse?

JN: Just as bad. Harvard was always, I think, at the cutting edge of the absurdity. Harvard was always the best at doing what was worst in our cultural environment. Harvard was the principal focus of scientism in philosophy. It was so depressing, you have no idea. For any young person who had a yearning for understanding, Harvard was a killer.

RB: Well, if this was so at Harvard, where was the philosopher at Iowa State or Wyoming Normal?

JN: Some of them might have been better, because they didn't have the money to get the most *scientistic* people. The point is, in the 1950s, when I was at the university, was the peak period when philosophy was *scientistic*. Hans Reichenbach in the early 1940s had redefined philosophy as a branch of science, and in the 1950s Harvard was always getting the best people in what was conventionally considered good. But Harvard was not a place to be if you were thirsty for the spirit. I don't know what it's like now, but in the 1950s it might have been better to be at Iowa State.

RB: So, speaking of today, young people in college are going to have to find it on their own.

JN: Absolutely. Now things are different. Harvard or Yale might be good. I don't know. The student has to smell out the professor who's got that something.

RB: She or he might not necessarily be in the philosophy department.

JN: No, she or he might be in the Egyptology department, might be in environmental science, might be in the infirmary, for all I know. The student is going to have to just use that sense of flair, of a feeling—and seek this somebody who has another dimension to them.

RB: Then students, as well as concerned individuals, have got to bring the question up—not only with themselves, but with the society in which they live. Start asking very tough questions of the scientist, the banker, the executive?

JN: As long as we are able to ask equally tough questions of ourselves. To do the first without the second is a recipe for continuing this madness. I have the right to ask you tough questions only to the extent that I can ask them of myself.

RB: In other words, you don't see salvation coming out of a movement—as salvation came out of the civil rights or the anti-Vietnam War protest. What about the current concern for ecology?

JN: I don't see it in any movements. I think the first thing you must say to yourself or to a person with this restlessness, is to try to become quiet in the self. Most of the evil in the world comes from an agitated mind, whether it's agitated about great truths or about falsehoods, about money or about God. Much of the evil in the world has come through agitation about God. Maybe now the evil comes through agitation about money, but it is the agitation that is the problem, the tension—not money or God. So people who blame religion or greed for wars are wrong. It's agitation, it's tension, it's fear. And the first thing you say to such a person is: Now try to collect yourself. Try to become quiet. Study the forces at work in yourself, and to the extent that you can really see them in *yourself,* you become a little freer. Only then can you study the strength of these forces in the world we're dealing with.

RB: In other words, the excessive scientific authority we're talking about, which helped lead us towards scientism, which has led to this mad materialism, is not necessarily going to be overwhelmed by opposition so much as it's going to be dissolved by being ignored.

JN: By being looked at dispassionately, compassionately. One needs to look at this scientism with dispassion and compassion, recognizing that I too have this impulse in me, to be too logical, to be computer-minded. That I too could, if you press a certain button in me, be the scientism-person myself. I've got it in me. It's not just out there in others, it's in all of us. Any movement is just going to add more agitation, going to perpetuate the stimulus-reaction chain which needs to be broken. And it can only be broken inwardly. It can't be broken externally in a mass way. I'm sorry, this may not be a satisfying answer, but these are the realities of the thing, I believe. For example, the first thing Hitler used to do with his audience, apparently, as far as I'm able to see, was to agitate. Once you get a mass of people agitated, you can do anything with them. Now, if you're going to get a people agitated with Eastern religion or with anything—

RB: —scientism?—

JN: —or with scientism, it's going to lead to madness. So media, newspapers, and all the forms of communication we have, have a general reaction on the mind of agitating the mind. You read the newspaper, and what do you get? Fear and outrage. Those are the two principal elements of everybody's daily morning diet. I get my coffee and my daily hit of fear and outrage.

RB: But then will individual non-attachment be effective in helping with haunting discontent?

JN: I think it will, if there are enough people. If there's a small nucleus of people who do that, yes. I think so. For the individual, yes, it would, if he works at that, because he will see that the discontent is not due to something external but is due to some arrangement of forces in himself which is not right. If I am in balance inwardly, nothing that happens outside is going to make me discontented.

RB: Is this Eastern fatalism?

JN: It's Western! It's Christian, it's Judaic. The principal message of all the spiritual tradition is "Seek ye first the kingdom of Heaven." That is, go inside. Know yourself. Socrates—"know thyself." That's where the stoics also began. Even the philosophy called Epicureanism begins with the idea of *ataraxia,* which refers to a movement of inner collectedness.

RB: So what you're saying is that many young people, even many older people, could come back from the ashrams and go back to Catholicism or Judaism or Protestantism.

JN: If Catholics and Jews and churches with this knowledge ever found their way to the real spiritual core of their own traditions, yes. But I don't know if they can do that. They may be looking, but so far, it's not too encouraging.

RB: Is this because religions seem so political?

JN: Perhaps, partly.

RB: Many who have agonized over the politics of the Moral Majority, over the extremism of Zionism, over the reactionary Pope or the number of closed churches here in America claim there's no spiritual foundation left.

JN: I think the churches—by that I mean the synagogues and churches—have been caught up in big social forces, and outer action has replaced inner search. Therefore, they're swept away in the same thing we are, and there's very little to find in them from the point of view we're speaking from. That doesn't mean it always will be so. Maybe a handful of Catholics or Jews will begin to find the life of their tradition again. In that case, it may nourish the tradition. There may be a rebirth. But I'm not optimistic about that scenario. I think the new path will come through science—that is to say, through a scientific language—because our culture has been conditioned by scientific thinking. So science is going to have to be placed in a new light. The discoveries of science are going to have to be placed in a broader metaphysical perspective so that we can see that what we've discovered really touches on a spiritual truth, and not just on some pragmatic thing—like mining the moon or something of that sort. And scientists are making discoveries all the time that really make you think we're on the verge of discovering an ancient metaphysical truth, but then they veer away from it, or else somebody comes in and excitedly says, "Wow!, this is mysticism!" Or "You're just doing what the Buddhists do!" It's neither of those. The discovery of how the brain works is one example. What goes on in the brain? Science is beginning to speculate that there are parts of the brain that are meant to channel higher forces. Science is beginning to touch on those things. They could be helped to see that.

RB: Are you saying there could be a scientific approach to this need for spiritual skill?

JN: Yes. But as I said in the beginning, you have to first separate ordinary science from inner work and see that those two reflect entirely different movements within the mind. Once you've seen that deeply enough, you can begin to see how they can move toward each other. But first you have to know that they're different, that the scientist as such, in his laboratory, no matter what he discovers, is not going to discover a metaphysical truth. He's using a part of the brain that is functionally incapable of seeing beauty and order and purpose in the universe.

RB: Where does he take all that's good about what he does? Does he have to leave the laboratory?

JN: He has to leave, or he has to bring another part of himself into the laboratory. That's a very good point, and we'd have to take a while to talk about it—how to be a spiritual seeker when you're in your laboratory.

RB: Do we need to differentiate between the brain and the mind? Some scientists claim the mind is in the brain, is merely a product of the brain, while the spiritualist claims the mind is outside the brain, beyond science and beyond the scientific claim that someday someone will prove that mind is produced by the brain.

JN: I think the mind *is* the brain. It's just that the brain is more spiritual than scientists think. I mean, there's nothing but physical reality in the world. Except that physical reality is permeated by metaphysical reality.

RB: But can we refine spiritual skill with science?

JN: No, but science, or the approach of science, the ideal of science, is noble—is that I wish to understand for myself without believing something just because it sounds good. If science could live up to that ideal, it would help us. But it doesn't.

RB: Why?

JN: Because the scientist is trapped in a part of his mind that he can't get out of. He is not objective.

RB: But the critic would say, why should he? That's not his role. Besides, he has the advantage of all this scientific authority. He is almost the prime minister of Western existence.

JN: He's miserable. He's suffering from all that so-called authority.

He knows he doesn't know. He knows he can't understand.

RB: And he knows he's responsible for nuclear waste.

JN: He knows it's all out of control, and he doesn't know what to do about it. Now the scientist is not objective. *That is the revolutionary notion.* The sin of scientism is that it's not objective enough. It's under the sway of all these emotions, all this subjectivity, all these cravings, all these predilections, all these short-sighted reactions. Am I putting this too provocatively? I'm not saying scientists should be less objective. I'm saying they should be more objective, which means they should be more in control of their own minds. That's what objectivity is, when you're in control of your mind, when you're a master of your own emotions and mind. The scientist is not the master of his emotions, therefore he cannot be objective. He can only be alienated. And alienation, which is cold, looks to the naive person like objectivity, but it is simply coldness. Objectivity is not coldness.

RB: Do you think he's alienated because of the specialization in science today?

JN: Yes, yes, specialization in the sense that—

RB: —it's been carried to an insane degree—

JN: —yes, but you need the movement toward specialization. And you need the movement toward integration, too.

RB: But you can't carry specialization to the point that one discipline can't speak to the other.

JN: That's right. But nobody speaks to anybody else. You see how hard it is for people, even us, to listen to each other! When I was working with physicians, doing my book on medicine, the first question I asked was "What are the non-scientific aspects of medicine? Are there any? Are there any non-scientific aspects of the practice of medicine?" Everyone said yes. "Are they important?" They said yes. Then I asked, "Are they in danger of being lost?" They'd say yes. They all agreed. "What is the single most important aspect of the practice of medicine that you might call the non-scientific side?" And they all agreed, almost unanimously, that it was listening. Until human beings can listen, they can never go beyond their trap. Listen to the universe, listen to the patient, listen to the self. The best doctors, the most searching ones, know that they don't really know how to listen. Listening—if I could put my finger on one thing that can get us out of this sin and scientism today, it's listening. We're all so specialized, we don't speak to each other. We can't communicate because we can't listen. We can't be open. We're not open.

RB: We're getting back to what you said much earlier about the role of the philosopher being to show people how to listen.

JN: Yes. You could say that if the philosopher could be the hero of listening, the one who knows how to be open to all the aspects of people in the world, then, yes, the philosopher could really help the scientist. In my experience, speaking to so many people, I've found that the only way to respond to another person humanly is to listen. It seems that we're always trying to say something before we hear the other person. And the physician, like the scientist, needs to be able to listen to the whole universe.

RB: This should be true of the executive or banker, as well as the concerned individual.

JN: I would say that. You know, it sounds like a panacea, it's very simplistic, but the secret of living is listening. This is not just a mental thing. It's a human act of attention. It means when I listen, another power in the mind is activated which is real attention, or an approach to real attention. And the secret of living is attention.

END

Jacob Needleman was interviewed in San Francisco, August 31, 1985.

BIBLIOGRAPHY

BATESON, GREGORY, *Steps to an Ecology of Mind.* Ballantine Books, New York, 1979.

BATESON, GREGORY, *Mind and Nature.* Bantam Books, New York, 1979.

BOYD, DOUG, *Rolling Thunder.* Delta Books, New York, 1976.

CAMPBELL, JOSEPH, *Hero with a 1,000 Faces.* Princeton University Press, 1968.

CAPRA, FRITJOF, *The Tao of Physics.* Shambhala, Berkeley, 1975.

FLEW, ANTONY, *A Dictionary of Philosophy.* St. Martin's Press, New York, 1979.

GREEN, ELMER & ALYCE, *Beyond Biofeedback.* Delta Books, New York, 1978.

MASLOW, ABRAHAM, *Toward a Psychology of Being.* Van Nostrand Reinhold, New York, 1968.

NEEDLEMAN, JACOB, *Consciousness & Tradition.* Crossroad Publishing, New York, 1982.

NEEDLEMAN, JACOB, *The Heart of Philosophy.* Bantam Books, New York, 1984.

NEEDLEMAN, JACOB, *The New Religions.* E.P. Dutton, New York, 1977.

NEEDLEMAN, JACOB, *A Sense of the Cosmos.* E.P. Dutton, New York, 1975.

NEEDLEMAN, JACOB, *The Way of the Physician.* Harper & Row, San Francisco, 1985.

PELLETIER, KENNETH R., *Toward a Science of Consciousness.* Celestial Arts, Berkeley, 1985.

ROSZAK, THEODORE, *Where the Wasteland Ends.* Doubleday, Garden City, New York, 1972.

SHELDRAKE, RUPERT, *A New Science of Life.* J.P. Tarcher, Los Angeles, 1983.

SMITH, HUSTON, *Forgotten Truth.* Harper & Row, New York, 1976.

THOMPSON, WILLIAM, *Passages About Earth.* Harper & Row, New York, 1981.

WILSON, COLIN, *Mysteries.* G.P. Putnam's Sons, New York, 1978.

YOUNG, ARTHUR, *The Reflexive Universe.* Robert Briggs Associates, San Francisco, 1984.

F R O M

Robert Briggs

A S S O C I A T E S

EAST & WEST by Stanislav Grof, M.D.

Heralded as one of the most comprehensive yet accessible overviews of the effects of ancient wisdom on modern science, EAST & WEST traces fundamental influences and reveals the principles and tenets involved.

32 pages $2.95 Broadside Edition

THE COMMON BOOK OF CONSCIOUSNESS by Diana Saltoon

Offering a simple yet comprehensive program for physical and psychological well-being, this book shows how to unify exercise, diet, meditation and lifestyle into four aspects of awareness that produce greater consciousness.

152 pages $7.95 Trade Paperback

THE BRAIN SCALE OF DR. BRUNLER by Arthur M. Young

Assessing the attempt of Oscar Brunler to measure brain radiation and, consequently, the evolution of the soul, Arthur Young points out how and why Brunler's ideas and efforts were important.

20 pages $2.95 Broadside Edition

DARWINISM by Norman Macbeth

Interviewed in Towards magazine, Norman Macbeth reveals why classical Darwinism has faded and deals with the irony that its passing—unknown to the public—is familiar to almost everyone working in the field of evolution.

32 pages $2.95 Broadside Edition

THE REFLEXIVE UNIVERSE by Arthur M. Young
Introduction by Jacob Needleman

Considered by Colin Wilson to be "the most important volume of cosmology to be published in the past twenty-five years," THE REFLEXIVE UNIVERSE integrates scientific laws with the evolution of consciousness, disclosing common ground between esoteric thought and modern science.

293 pages $12.95 Trade Paperback

THE FOUNDATIONS OF SCIENCE by Arthur M. Young

THE FOUNDATIONS OF SCIENCE summarizes the thesis convincingly spelled out in Arthur Young's THE REFLEXIVE UNIVERSE and THE GEOMETRY OF MEANING. Examining contributions from Planck and Heisenberg, from Bateson and Bohm and Szent-Gyorgyi, he provides an overview of the effects of quantum physics and the consciousness movement on the history of science.

26 pages $3.95 Broadside Edition

A NEW AGE by Kenneth R. Pelletier

In this unusal interview Kenneth Pelletier talks about the demands and promises of a new age. Present and future health care systems are discussed, along with contemporary doubts about Freud, the brain/mind problem and the astonishing growth of research in psycho-immunology.

44 pages $3.95 Broadside Edition

Order Form

To order: enclose check or money order and mail to:
Robert Briggs Associates
Publishers Services
POB 2510
Novato, California 94948

Qty	Title	Price	Amount
________	THE REFLEXIVE UNIVERSE	@ $12.95 (paper-6-9)	________
________	BRAIN SCALE OF DR. BRUNLER	@ 2.95 (paper-01-7)	________
________	EAST & WEST	@ 2.95 (paper-00-9)	________
________	COMMON BOOK OF CONSCIOUSNESS	@ 7.95 (paper-3-4)	________
________	DARWINISM	@ 2.95 (paper-8-5)	________
________	FOUNDATIONS OF SCIENCE	@ 3.95 (paper-03-3)	________
________	A NEW AGE	@ 3.95 (paper-02-5)	________
________	SIN & SCIENTISM	@ 3.95 (paper--7-7)	________

Purchase ________

California sales tax 6%

In BART Counties, Santa Clara, Santa Cruz,
San Mateo, & Los Angeles Counties add 6½% ________

Postage & Handling:
Orders UNDER $6 - add 1.00
Orders OVER $6 - add 2.00 ________

TOTAL AMOUNT ________

☐ CHECK ENCLOSED ☐ VISA ☐ MASTER CHARGE

ACCT #: ________________________________

EXP DATE: ______/__________ SIGNATURE:________________________

____

SHIP TO:

____

ADDRESS:

____

CITY

____

STATE ZIP